LET'S EAT BUGS!

A Thought-Provoking Introduction
to Edible Insects
for Adventurous Teens and Adults

(2nd Edition)

by

MK Grassi

MKClassicBooks.net

MK Classic Books

LET'S EAT BUGS!
A Thought-Provoking Introduction to Edible Insects for Adventurous Teens and Adults
(2nd Edition)

Copyright © 2014 Richard Grassi & GO International Enterprises, Inc.

All Rights Reserved. No part of this book may be reproduced, copied, stored, or transmitted in any form or by any means (graphic, electronic or mechanical, including photocopying, recording, or in information storage and retrieval systems) without the prior written permission of the author.

Published by GO International Enterprises, Inc. and MK Classic Books.
For more information about MK Classic Books, please visit:
http://mkclassicbooks.net/.
Contact: MKClassicBooks@gmail.com

Images and Photographs: Royalty free images reproduced under license from stock image repositories. Bee photo on page 12 © Cowboy54 | Dreamstime.com - Stingless Bee (Trigona Pagdeni).

WARNING
The material in this book is intended to be of general informational use and does not constitute dietary or medical advice. The author recommends caution and careful research. Please check with a physician before eating any kind of new, unusual or unfamiliar food.

NOTICE OF LIABILITY
In no event shall the author or the publisher be responsible or liable for any loss of health, profits or other commercial or personal damages, including but not limited to special incidental, consequential, or any other damages, in connection with or arising out of furnishing, performance or use of this book.

This book is dedicated to my brother, Daniel Grassi, and his wonderful spirit of adventure.

Table of Contents

Buggy Introduction	*7*
Help End World Hunger	*8*
Greener World	*9*
Business Opportunity	*10*
Which Bugs Can You Eat?	*11*
Ants	*13*
Bees	*15*
Beetles	*17*
Caterpillars	*19*
Cockroaches	*20*
Crickets	*21*
Grasshoppers	*22*
Hornworms	*23*
June Bugs	*24*
Locusts	*25*
Millipedes	*26*
Scorpions	*27*
Tarantulas	*29*
Termites	*30*
Wasps	*31*
A Comment About Insect Recipes	*33*
Insect Recipe Resources	*34*
Considerations	*35*
Catching, Raising or Buying	*42*
Recommended Resources	*43*
Buggy Conclusion	*45*
Recommended General Resources	*50*

Buggy Introduction

Some of us avoid bugs* whenever possible, so the idea of eating them may seem pretty gross. But believe it or not, about two billion people do just that.

Grasshoppers, ants, butterflies, scorpions and even some types of cockroaches are quite nutritious, and they can be deep-fried, stir-fried, grilled, baked, boiled or barbequed. You can even safely eat certain types of tarantulas and centipedes.

The United Nations said in a recent report that more people should eat bugs. (A link to that report, which can be downloaded for free, is provided in the Recommended Reading section of this book.)

No, that statement was not made on April Fools' Day, and the people who are promoting bug eating did not time travel from the Ice Age!

Actually, there are some very good reasons why these researchers think eating bugs is a smart idea.

Here are three of them...

*Technically, the terms 'bug' and 'insect' are not the same. But in this book, they are used interchangeably (and rather loosely) to include a variety of creepy crawlers. Basically, all bugs are insects, but not all insects are bugs. A true bug belongs to the order Hemiptera, which includes cicadas, assassin bugs, water bugs and aphids among others.

Reason 1 -- **Help End World Hunger**

Despite the amazing technological advances of the past century, billions of people still do not have enough nutritious food. While some of us can simply go to a supermarket and buy beef, chicken, pork, cheese and other foods, a lot of other people do not have enough to eat and are suffering from malnutrition.

However, many of these same people could easily find bugs, as there are many more insects than people in the world, and bugs live in every country, town and village on earth. Because bugs have nutritional value, they could become a large part of the answer to the problem of world hunger.

Reason 2 -- **Greener World**

Eating bugs is more environmentally friendly than eating other types of meat. Another United Nations report[1] says that animal agriculture is a major cause of deforestation, water and soil problems, air pollution and the reduction of biodiversity.

And conventional livestock (cows, pigs, sheep, etc.) produces more greenhouse gas than cars and trucks. Thus, if a lot more of us were to eat bugs, the world environment would be better off.

---

[1] Henning Steinfeld, Pierre Gerber, Tom Wassenaar, Vincent Castel, Mauricio Rosales, and Cees de Haan, "Livestock's Long Shadow: Environmental Issues and Options" (2006), Food and Agriculture Organization of the United Nations.

Reason 3 – **Business Opportunity**

Insect cultivation and sales could become a profitable and accessible line of work for many people.

Bug eating on a large scale would require a large supply of insects. A large supply of insects, in turn, would imply the establishment of a large 'bug industry'.

Since it does not require a lot of money to learn about and capture or cultivate insects, spiders, scorpions, centipedes, etc, a big bug industry could provide much needed jobs and business opportunities for people around the world.

Which Bugs Can You Eat?

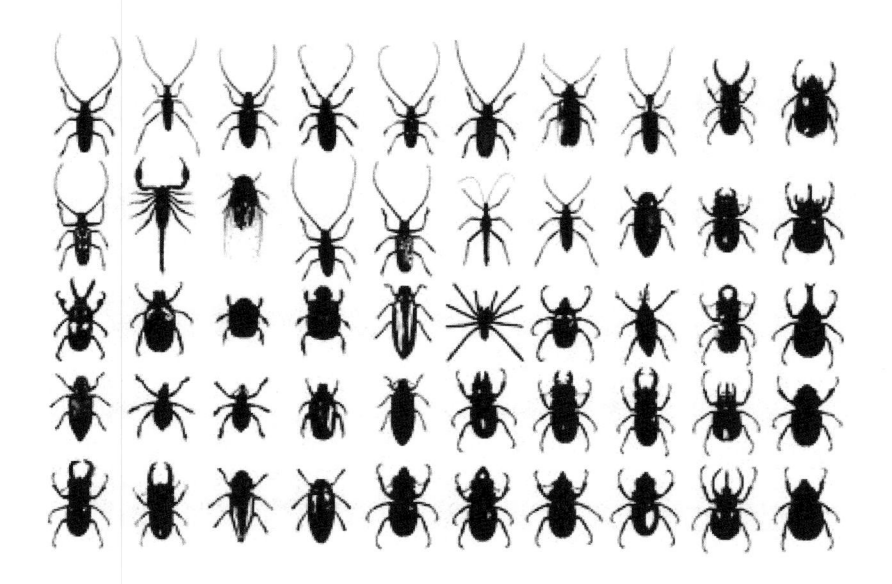

Answer: There are so many different types of edible insects that you are sure to like at least a few of them if you are brave enough to try.

In fact, scientists estimate that there are probably at least 1900 different types of bugs that can be safely eaten.

Let's take a look at fifteen different creepy crawlers that could actually make a safe and tasty addition to your diet.

Ants

There are dozens of ant species in the world that you can eat, and each type has its own unique flavor. Leafcutter ants are prepared in the same way as popcorn and they taste a bit like a mix between bacon and pistachio nuts.

Weaver Ants can be eaten raw or cooked (baked, fried or sautéed) and they taste like shrimp, while Flying Ants taste like buttery pork rinds and are usually roasted with salt and lemon juice.

Bees

Adult bees are usually roasted and then turned into a type of flour, which is then used in the same way as normal flour.

Bee larvae, on the other hand, are quite tasty on their own. They can be stir-fried, deep fried or baked. And depending on what spices you put on them, they can taste like nuts or mushrooms with bacon. In Mexico, some people cover baked bee larvae with chocolate and then sell them as popular snacks.

Beetles

Beetles taste like nuts and both adult beetles and beetle larvae can be eaten.

Dung beetles are quite popular in several countries in Asia, although you can only safely eat the ones that are grown commercially, as wild dung beetles have a less than healthy diet and can make you sick. These beetles are usually cleaned before they are fried together with vegetables and/or meat such as pork.

You can also eat Rhino Beetles and Phyllophaga Rugosa Beetles. These beetles are quite popular in Thailand and are usually boiled, mixed with lemon grass, lime leaf, galigale and garlic, salt and soy sauce and then dehydrated and bagged for future consumption. Dung beetles can be prepared in this fashion as well.

Caterpillars

tent caterpillars

There are many kinds of edible caterpillars, and they often have a fruity taste. They are generally boiled, left to dry in the sun and then eaten.

Popular species include sapelli, tent and bamboo caterpillars. Caterpillar dishes are especially popular in sub-Saharan Africa.

19

Cockroaches

You cannot eat the cockroaches you find around your home or out in the yard, but these small creatures are edible if they are raised on a diet of fresh fruits and vegetables.

They actually taste a bit like greasy chicken and can be toasted, fried, sautéed or boiled.

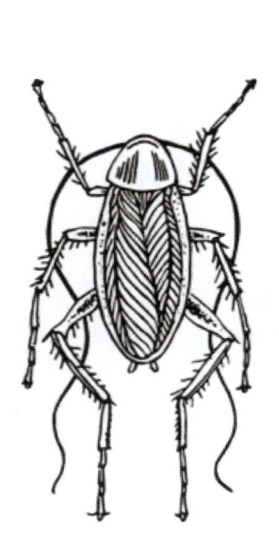

Depending on how they are cooked (and your own preference), you can eat cockroaches with ketchup, mustard or soy sauce.

Crickets

Crickets are one of the most popular edible insects in the world. They can be fried, sautéed, boiled or roasted.

Flavor varies depending on how they are cooked; some people say they taste a bit like cockroaches while others say they taste like nuts. Mole crickets are usually fried without flavoring and taste a bit like liver. They are eaten on their own and are considered to be a delicacy by many insect lovers.

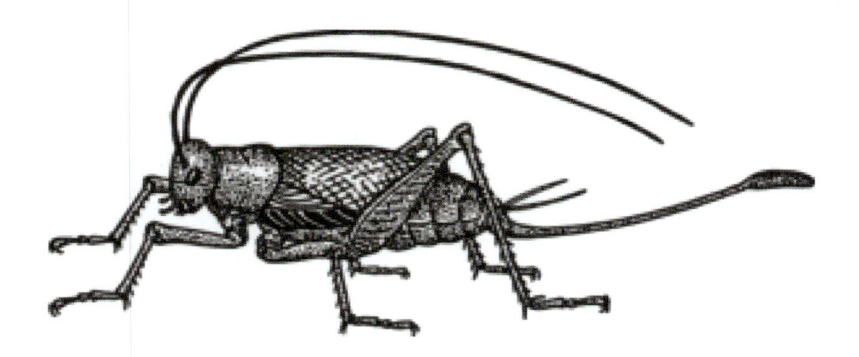

(Crickets are favorably cited in the Bible; see *Leviticus* 11:21 & 11:22.)

Grasshoppers

You can eat any grasshopper that is solid colored but you should avoid the multicolored ones as they are toxic. Roasted grasshoppers in Mexico are usually eaten with chili and lime and have a sour, spicy taste.

It is also possible to eat grasshoppers raw, although most people prefer to fry or sauté them either on their own or with a few crickets added. Fried and sautéed grasshoppers taste a bit like crunchy chicken.

Hornworms

Tomato Hornworms and Tobacco Hornworms are edible, but they usually eat plants that are toxic to humans. So, you need to feed hornworms green peppers, tomatoes or ready-made hornworm chow for a few days before you can cook them.

Hornworms are usually fried and taste like a combination of green tomatoes, shrimp and crab.

June Bugs

You can eat both adult June Bugs and its larvae. Unlike many other insects, June Bugs do not eat plants that are harmful to humans so all you have to do is catch them and cook them.

June Bugs can be added to mulberry sauce but they are usually fried, prepared in the same way as popcorn or roasted. Just be sure to remove the wings and legs before eating them and then sit back and enjoy.

June Bugs taste like walnuts with a bit of a buttery flavor.

Locusts

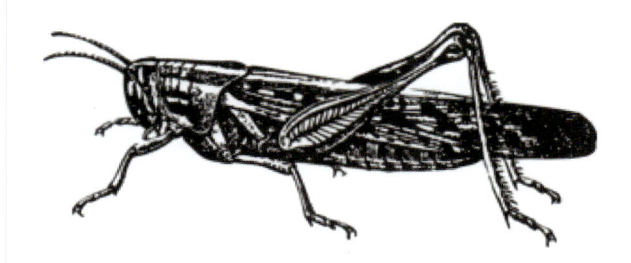

Locusts have a vegetable type flavor, although it is hard to pinpoint which exact vegetable they taste like. These jumpy insects are usually boiled in salt water and then set out to dry, although they can also be eaten without drying.

Conversely, some people simply put the locusts out to dry and then eat them raw. The wings and legs should be removed whether you cook the locusts or not.

Millipedes

Millipedes are a delicacy but they are not easy to prepare, as their toxic glands must be carefully removed. Also, millipedes can be quite costly; however, they are well worth the expense.

It is not hard to find grilled, skewered or fried millipedes for sale in Asia. They taste like chicken and can either be eaten plain or with a bit of ketchup or soy sauce.

Scorpions

Scorpions can be skewered on sticks or fried and eaten with rice and/or other dishes. They taste like either soft shell crab or crunchy crab, depending on how you cook them.

Scorpions can be eaten whole, as cooking them destroys the poison found in the stinger; however, you are not missing out on anything if you cut a scorpion's tail off before preparing it.

water scorpion

Water scorpions are also edible, but you have to remove the wings first.

You can either fry them or dry them out and preserve them.

Please note, however, water scorpions tend to taste much better fresh than they do out of a bag or can.

Tarantulas

Tarantulas became popular in Cambodia in the 1970s when many people were facing starvation during Pol Pot's tyrannical rule. And these big spiders are still popular today, even among those who are well-off and can afford to eat other foods.

Cambodians cook tarantulas with salt, oil, sugar and garlic. The legs taste a bit like crab and have a crunchy texture; the stomach is soft and has a nutty flavor. Generally speaking, people either really like this large hairy spider or totally hate it; there seems to be no neutral ground.

Termites

Termites have a slightly minty flavor but other than that do not have much of a taste of their own. They are actually very good for you as they are rich in the amino acids that your body needs in order to stay healthy and in good shape.

Termites can be eaten raw, but they are usually roasted or fried. The wings must be removed before you eat them and you will probably want to add a bit of salt or a topping such as soy sauce or ketchup.

Queen termites are prized delicacies among bug connoisseurs.

Wasps

There are many ways in which you can eat wasps. They can be sautéed, roasted, fried or boiled and they have a buttery yet earthy taste.

Both adult wasps and wasp larvae are considered safe to eat, although most people find that wasp larvae are tastier and easier to prepare than adult wasps.

Wasp larva ("hachi-no-ko") is a traditional, but nowadays rarely eaten, food in some mountainous regions of Japan where it is cooked with soy sauce and sugar.

There are two main types of wasps; solitary wasps and social wasps. Both kinds can be eaten but the latter is more common.

In Mexico, social wasps are captured along with their paper wasp nests and the whole nest is roasted.

Wasps Storage
Wasps can also be cooked, dehydrated and then stored away to be eaten another day. If a bag of properly prepared wasps is tightly sealed, the wasps will stay good for about a year.

A Comment About Insect Recipes

This publication is not a recipe book; it is a concise and hopefully entertaining introduction to entomphagy. However, a sample recipe has been included* here to provide an example of what you can expect to find if you search for insect recipes online and elsewhere. (Please see the next page for a list of recommended resources.)

Bee-LT Sandwich

¼ cup Bee larvae
1 egg white
1 tsp butter
1/4 tsp honey
1 tomato
1 leaf lettuce
2 slices of bread
1 tbsp mayonnaise
1 pinch salt

Sauté the bee larvae, salt and honey in the butter
until it is golden brown and crispy.
Remove and mix in egg whites.
Return it to the hot butter and mash it together into a patty.

Toast bread and slice tomato.
Spread mayonnaise on toasted bread when ready.
When bee patty becomes firm, place it atop
the lettuce and tomato on the sandwich.

NOTE: This recipe and several others can be found at Daniella Martin's helpful blog, 'Girl Meets Bug' (http://edibug.wordpress.com/recipes/). Girl Meets Bug is a great resource, and I highly recommend it along with Ms Martin's book, *Edible: An Adventure into the World of Eating Insects and the Last Great Hope to Save the Planet.*

*Reprinted with the author's permission.

Insect Recipe Resources

Online
http://insectsarefood.com/recipes.html (approx 20 recipes!)

http://edibug.wordpress.com/recipes/
(blog name: "Girl Meets Bug" - fantastic resource!)

http://www.ent.iastate.edu/misc/insectsasfood.html

http://frogsonice.com/froggy/recipes.shtml

http://coolbugstuff.com/recepies.php

Videos
youtube.com/watch?v=_bGKQsP9hpQ   (Thai cricket recipe)

youtube.com/watch?v=3f7I_HAm4d8

youtube.com/watch?v=B0hv1j1KRfk

youtube.com/watch?v=OoEIK593epY#aid=P-FNBrqHH4Y

Books
Top 50 Most Delicious Insect Recipes by Julie Hatfield

Edible: An Adventure into the World of Eating Insects and the Last Great Hope to Save the Planet by Daniella Martin

Eating Insects, Eating Insects As Food by Elliot Lang

NOTE: More recipe suggestions can be found in the 'Recommended General Resources' section at the end of this book.

Considerations

Every proposal comes with its own set of challenges and objections. What are some of the real or imagined drawbacks of eating bugs?

Here are three commonly cited concerns...

Concern 1 – Illness

Q. Can't we get sick from eating bugs?

A. Yes, it's possible, but parasitic infections from eating hard-shelled insects like beetles, even uncooked ones, are very rare. Just be sure you thoroughly cook your bugs.

Q. What about poisons? Some bugs are dangerous!

A. That's true. And that's why it's best to check to see if your insect is listed as safe to eat with a trusted source like the United Nation's Food and Agricultural Agency (a link to their website is provided in the "Recommended Reading" section of this book.)

The trick to eating and enjoying any insect is to cook it well. Even if a bug has venom or toxins, a thorough boiling will usually neutralize the bad stuff. Besides making insects safe to eat, cooking them also improves the taste.

Q. But what about chemical toxins?

A. That's a valid concern too.
The majority of insects that people eat are wild-harvested, so it's difficult to know what the insects themselves have been exposed to. And this is a serious issue because most farming utilizes lots of pesticides and other toxic chemicals.

Where insects live and what they eat is very important. You should never eat raw insects unless they've been bred and raised by a trusted source because it's difficult to determine if a raw insect has been tainted with pesticides.

If, for example, a bug recently ate plants sprayed with pesticide or herbicide, those dangerous chemicals are now inside its body. To solve this problem, just feed it fresh organic plants for 24 hours (for purging) and/or cook it thoroughly.

Also, it's better to stick with live insects if you don't know what has killed the dead ones. And you can kill the insects simply by putting them in your freezer.

Q. What about shelf life? How long can we safely keep insects before they spoil from bacteria?

A. While it's true that spore forming bacteria can be a concern even for cooked and commonly eaten insects such as crickets and meal-worms, we can greatly reduce the likelihood of spoilage by using simple preservation methods like drying and acidifying (without refrigeration).

Keep in mind that when it comes to preparing insects, we don't have to start from scratch; we have thousands of years of traditional cooking methods to use and build upon.

Concern 2 – **Allergies**

Q. I'm squeamish around bugs. Maybe I'm allergic to them. So the thought of eating them and feeling their little legs in my mouth is totally gross!

A. That's not a question! LOL. Feeling disgusted is not the same as being allergic... Remember, you can remove the legs and head of an insect before eating it if you want.

Although people with no history of arthropod sensitivity are unlikely to experience allergic reactions to insects, if you have problems with shellfish or shrimp or even chocolate, it's probably best to avoid eating insects, until you figure out your tolerance levels.

Concern 3 – **Too Small, Still Hungry**

termite entrance

Q. Bugs are very small. How can we breed enough of them to produce significant amounts of food? It appears that most countries where insects are a staple food still suffer from hunger!

A. The problem of hunger in poor and developing countries is frequently accompanied by corrupt or unstable governments and poorly developed infrastructures, which badly affect any kind of agricultural programs including insect farming.

Nevertheless, some African, Latin American and Asian countries (like Thailand) already have commercial insect farms and more are being built. But, for better or worse, at present 'insects as food' is mainly a small-farm production industry.

In order to develop this industry on a global scale, there needs to be better coordination in the areas of production, distribution, marketing and sales. And governmental regulations and laws may need to be drawn up and enforced to avoid problems down the road. (The U.S. government, for example, currently doesn't regulate insect production for food safety.)

Catching, Raising or Buying Edible Insects: Which is Best?

This question can only be answered by you because it depends on your own personal circumstances. Your location, finances, health and physical mobility... all need to be considered in order to determine which option is most suitable for you.

Needless to say, catching your own insects is the cheapest option in terms of money, but you may also find it to be the most time-consuming and physically tiring choice.

Raising insects (home-grown or via store bought kits) enables you to have a steady supply of fresh food that you do not have to chase or hunt down. Plus, you can control what chemicals your bugs are being exposed to. Of course, you still have to invest time, money and space on their upkeep, so they are not a free meal like gathered insects.

Shopping is arguably the most expensive choice for procuring edible insects, but in some ways, it is also the easiest. If you live in Asia, Latin America or Africa, you may be able to get inexpensive bugs, fresh or already cooked, at local markets or food stalls. If, on the other hand, you live in a western country, your best option is likely going to be a vendor who is able to ship insects to your home.

On the next page is a list of recommended resources for each option. Please be aware that some of the information, particularly in regard to shop or vendor names, addresses, phone numbers, etc, can change rapidly.

Recommended Resources for Catching, Raising and Buying Edible Insects

Online Resources – Catching & Identifying Edible Bugs
https://www.youtube.com/watch?v=VTKdm6kGy4U (video)

http://www.motherearthnews.com/real-food/edible-insects-zebz1305znsp.aspx

http://adventure.howstuffworks.com/survival/wilderness/edible-bug1.htm

Online Resources – Bug Kits & Raising Insects
http://www.openbugfarm.com/

http://www.tiny-farms.com/blog.html

https://www.youtube.com/watch?v=WGdBEWxmJp0 (video)

https://www.youtube.com/watch?v=rHsyaucEQFw (video)

Books – Raising Insects
Cricket Breeding Made Easy: Your Guide to Raising Healthy Feeder Crickets
by JM Daniels

Raising Mealworms 1-2-3: How to Breed and Raise the Easiest Feeder Insect By Life Cycle
by JM Daniels

The Bee Book For Beginners: 2nd Edition (Revised) : An Apiculture Starter or How To Be A Backyard Beekeeper And Harvest Honey From Your Own Bee Hives
by Frank Randall

(Resources continued on next page.)

Resources for Buying Edible Insects

Possibly the best place to visit online is 'Girl Meets Bug', specifically: http://edibug.wordpress.com/where-to-get-bugs/. This is a popular gathering spot for hobbyists, researchers, bug food connoisseurs and professional insect farmers. As such, the information on the site, especially in the readers' comments, is up-to-date and helpful.

USA

Rainbow Mealworms
126 East Spruce ST
Compton, CA
90220 USA
tel. 310-635-1494
www.rainbowmealworms.net

Fluker's Farms
P.O. Box 530
Port Allen, LA
70767   USA
tel. 800-343-8537
www.flukersfarms.com

UK (including Europe, with worldwide shipping)

Bug Grub Ltd
4 Lynn Road
King's Lynn
Norfolk
PE33 0EW
United Kingdom
tel. 44(0)5603 671159
www.buggrub.com

Australia

www.ediblebugshop.com.au
No postal address online
Email: bug@ediblebugshop.com.au
tel. 0431-533-955

Buggy Conclusion

Although it is true that there is not much demand right now for edible insects in Western Europe, Australia, New Zealand, Canada and the United States, there is a long-established appreciation for tasty bugs in other parts of the world, especially Asia.

The fact is many insects, spiders and worms are not only edible but also very good for us. They can provide protein, iron, amino acids and numerous vitamins and minerals.

Moreover, 80% of the world's cultures already eat insects, and in some places certain types of bugs are highly esteemed delicacies.

Most edible insects do not cost a lot of money; in fact, in some cases you can get these yummy little creatures for free if you live in a rural area or in a house with a back or front yard.

However, if bug hunting is not your cup of tea, then you can simply buy the insects online.

As we have already discussed, beetles and wasps (but also other bugs such as crickets, termites, Sago Worm larvae, centipedes, ant eggs and scorpions) can be preserved and shipped to your front door.

If you live in Asia, you may be able to visit a restaurant or street vendor that sells cooked bugs and choose the species and style of cooking that you like best.

On the other hand, if you are feeling adventurous, then do not hesitate to catch your own food and prepare it as you see fit — with adult supervision, of course.

If you are a newbie insect eater, use a cookbook or recipe (either of which are easy to find on the internet) to make sure that your bugs are being properly prepared, as some insects contain toxic portions that should not be eaten.

With a little bit of research and experience, you can enjoy a nutritious, tasty bug meal if you are willing to put aside pre-conceived ideas. Just sample a few bugs and see which ones you like best. Then you can start expanding your menu over time with other insects.

Let's eat bugs!

* * * * *

Hi, I hope you have enjoyed this book and have discovered some interesting ideas! If so, please add a customer review of ***Let's Eat Bugs!*** at Amazon. Your thoughts are very important to me. Thanks!

MK Grassi

P.S. Info about my other books and a fun, free game called "The Big Bug Word Search Puzzle" are available at **MKClassicBooks.net**.

P.S.S. Please go to the next page to see Recommended General Resources.

Recommended General Resources

Books (available at Amazon)

Man Eating Bugs: The Art and Science of Eating Insects
by Peter Menzel and Faith D'Aluisio

Eat-a-bug Cookbook, Revised: 40 Ways to Cook Crickets, Grasshoppers, Ants, Water Bugs, Spiders, Centipedes, and Their Kin
by David George Gordon

Creepy Crawly Cuisine: The Gourmet Guide to Edible Insects
by Julieta Ramos-Elorduy

The Insects: An Outline of Entomology
by P.J. Gullan and P.S. Cranston

Borror and DeLong's Introduction to the Study of Insects
by Norman F. Johnson and Charles A. Triplehorn

How To Know Insects by Roger G. Bland and H.E. Jaques

Edible Insects: Future Prospect for Food and Feed Security (Fao Forestry Paper) by Food and Agriculture Organization of the United Nations

The Insect Cookbook: Food for a Sustainable Planet (Arts and Traditions of the Table: Perspectives on Culinary History)
by Arnold van Huis, Henk van Gurp, Marcel Dicke, Françoise Takken-Kaminker, and Diane Blumenfeld-Schaap

Videos

"Life in the Undergrowth" (available at Amazon)

"Life Season 1, Ep. 6 – Insects" (available at Amazon)

YouTube Videos
There are many helpful videos on YouTube, and more are being added every day. Just do a search and you will find plenty! Below are some of the more interesting ones I've seen.

YouTube: "Which Bugs Are Safest to Eat -- and Tastiest?"
(http://www.youtube.com/watch?v=dD6AG3HWwCk)

YouTube: "Girl Meets Bug - ep.1 - WaxWorm Tacos"
(http://www.youtube.com/watch?feature=player_embedded&v=fA_rBNeVtzo)

YouTube: "Can Eating Insects Save the World – BBC"
(http://www.youtube.com/watch?v=Acxbx-DUkL4)

YouTube: Why not eat insects? - TED TALKS
(http://www.youtube.com/watch?v=pRoUQ2N5nVk)

YouTube: "Future of Meat: Edible Bugs As Cheap..."
(https://www. youtube.com/watch?v=KYhN_qms474)

YouTube: "Insect Meat: Food Of The Future"
(https://www.youtube.com/watch?v=1YrmNeGPMgU)

YouTube: "How To Eat Insects, Worms and Bugs!"
(https://www.youtube.com/watch?v=7_4avZiKD2o)

YouTube: "Edible Insects: Finger-lickin' Grub – KQED Quest"
(https://www.youtube.com/watch?v=hgqZe8Gn9Oc)

Online Resources

United Nation's Food and Agricultural Agency (FAO)
(http://www.fao.org/forestry/edibleinsects/en/)

Great free report by the FAO
(http://www.fao.org/docrep/018/i3253e/i3253e00.htm)

Girl Meets Bug (very entertaining and informative!)
(http://girlmeetsbug.com)

The Atlantic – Can We End Hunger By Eating Bugs?
(http://www.theatlantic.com/international/archive/2013/05/can-we-end-hunger-by-eating-bugs/275997/)

National Geographic: Bugs as Food: Humans Bite Back
(http://news.nationalgeographic.com/news/2004/04/0416_040416_eatingcicadas.html)

Business Insider -- We Can Make Insects Taste Like Buttery Popcorn
(http://www.businessinsider.com/list-of-edible-insects-2013-5?op=1)

A Concise Summary of the General Nutritional Value of Insects
(http://www.food-insects.com/Insects%20as%20Human%20Food.htm)

BBC -- Eating Insects: Would you cook with grubs?
(http://www.bbc.co.uk/food/0/21260185)

The Guardian – Insects Could Be The Key…
(http://www.theguardian.com/environment/2010/aug/01/insects-food-emissions)

PBS – Bugs for Dinner
(http://www.pbs.org/newshour/rundown/2012/05/bugs-for-dinner.html)

Notes

Printed in Great Britain
by Amazon.co.uk, Ltd.,
Marston Gate.